European Military Prospects, Economic Constraints, and the Rapid Reaction Force

Charles Wolf, Jr.

Benjamin Zycher

Prepared for the Office of the Secretary of Defense
and the Smith Richardson Foundation

NATIONAL DEFENSE RESEARCH INSTITUTE

RAND

The research described in this report was sponsored by the Office of the Secretary of Defense (OSD). The research was conducted in RAND's National Defense Research Institute, a federally funded research and development center supported by the OSD, the Joint Staff, the unified commands, and the defense agencies under Contract DASW01-01-C-0004.

Library of Congress Cataloging-in-Publication Data

Wolf, Charles, 1924-
European military prospects, economic constraints, and the rapid reaction force / Charles Wolf, Jr., Benjamin Zycher.
p. cm.
MR-1416
Includes bibliographical references.
ISBN 0-8330-3070-1
1. Europe—Defenses. 2. European Union. 3. Europe—Defenses—Economic aspects. I. Zycher, Benjamin. II. Title.

UA646 .W65 2001
355'.03304—dc21

2001048554

Cover design by Peter Soriano

Published 2001 by RAND
1700 Main Street, P.O. Box 2138, Santa Monica, CA 90407-2138
1200 South Hayes Street, Arlington, VA 22202-5050
201 North Craig Street, Suite 102, Pittsburgh, PA 15213
RAND URL: http://www.rand.org/
To order RAND documents or to obtain additional information, contact Distribution Services: Telephone: (310) 451-7002;
Fax: (310) 451-6915; Email: order@rand.org

RAND 1700 Main Street, PO Box 2138 Santa Monica, CA 90407-2138

ERRATA

PUBLICATIONS DEPARTMENT

April 1, 2002

To: Recipients of *European Military Prospects, Economic Constraints, and the Rapid Reaction Force*, by Charles Wolf, Jr., and Benjamin Zycher (MR-1416-OSD/SRF)

From: Publications Department

Begin at page iii. Replace text beginning with the second paragraph with the following:

The project included a workshop held at the Netherlands Institute of International Affairs at Clingendael in The Hague at the end of February 2001, attended by several experts from Germany, the United Kingdom, France, Italy, the Netherlands, and the United States. The workshop was organized to elicit the views and additional data from the European participants, who included Professor Keith Hartley and Lord John Roper from the United Kingdom, Dr. Uwe Nerlich and General Dr. Klaus Wittmann from Germany, M. Gilles Andreani from France, Dr. Stephano Silvestri from Italy, and Dr. Alfred van Staden, Professor Rob de Wijk, Dr. Hans La Bohm, and Colonel Franz Osinga from the Netherlands. The value of the workshop was significantly enhanced by the participation and assistance of David Gompert, president of RAND Europe, and Stephan de Spiegeleire, a member of the RAND Europe staff.

This report has benefited from comments on an earlier draft made by RAND colleagues Robert Hunter (formerly U.S. Ambassador to NATO) and James Thomson, and by Colonel Antoine Jaureguiberry of the French Army.

Replace all text on page iv with the following:

It is a pleasure for the authors to acknowledge and express appreciation for the valuable assistance provided by these individuals and institutions, while at the same time absolving them of any responsibility for the content of this report.

The project was executed through RAND's National Defense Research Institute (NDRI), a federally funded research and development center sponsored by the Office of the Secretary of Defense, the Joint Staff, the unified commands, and the defense agencies.

The report should be of interest and use to those in the policy community concerned with NATO, relations between the European Union and NATO, and the outlook for economic growth and military spending in Germany, France, the United Kingdom, and Italy.

Replace Figure B.2 at the top of page 48 with the following figure:

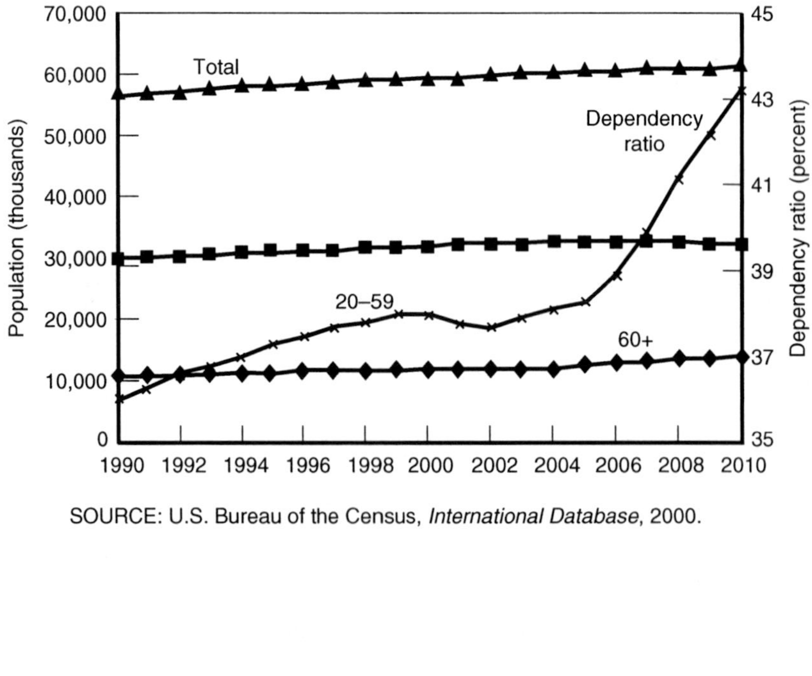

SOURCE: U.S. Bureau of the Census, *International Database*, 2000.

FOREWORD

This report was produced under a project jointly sponsored by the Office of Net Assessment in the Department of Defense and the Smith Richardson Foundation. Their sponsorship was based on an understanding that their joint support would enable the work to be expanded beyond what would have been possible if funding were confined to one sponsor alone.

The project included a workshop held at the Netherlands Institute of International Affairs at Clingendael in The Hague at the end of February 2001, attended by several experts from Germany, the United Kingdom, France, and Italy. The workshop was organized to elicit the views and additional data from the European participants, who included Professor Keith Hartley and Lord John Roper from the United Kingdom, Dr. Uwe Nerlich and General Dr. Klaus Wittmann from Germany, M. Gilles Andreani from France, Dr. Stephano Silvestri from Italy, and Dr. Alfred van Staden, Professor Rob de Wijk, Dr. Hans La Bohm, and Colonel Franz Osinga from the Netherlands.

This report has benefited from comments on an earlier draft made by RAND colleagues Robert Hunter (formerly U.S. Ambassador to NATO) and James Thomson, and by Colonel Antoine Jaureguiberry of the French Army.

It is a pleasure for the authors to acknowledge and express appreciation for the valuable assistance provided by these individuals and institutions, while at the same time absolving them of any responsibility for the content of this report.

The project was executed through RAND's National Defense Research Institute (NDRI), a federally funded research and development center sponsored by the Office of the Secretary of Defense, the Joint Staff, the unified commands, and the defense agencies.

The report should be of interest and use to those in the policy community concerned with NATO, relations between the European Union and NATO, and the outlook for economic growth and military spending in Germany, France, the United Kingdom, and Italy.

CONTENTS

FIGURES

TABLES

SUMMARY

This study of European military prospects and economic constraints (EMPEC) was motivated by the long-standing concern of U.S. policymakers with moving toward a better balance with our European allies in providing military capabilities, as well as in the sharing of responsibilities, decisionmaking, and costs of our alliance relationship. This concern is largely shared by our NATO allies, who also nurture—with varying degrees of intensity—the additional aim of enhancing the autonomous military capabilities of the European Union (EU) apart from the NATO alliance, while at the same time moving toward a better balance within the alliance itself.

The principal focus of the EMPEC project is the Union's European Security and Defense Policy (ESDP) and the ESDP's instrument, the Rapid Reaction Force (RRF), with secondary attention to the Defense Capability Initiative (DCI) within NATO and the relationship between the ESDP and the DCI. The DCI is a separate set of commitments by the members of NATO to upgrade their individual and collective capabilities and thereby enhance their capacity for bearing a larger share of the burden of and responsibility for NATO operations, together with U.S. forces. While these issues have many important political, diplomatic, and legal aspects, the narrower focus of this study is on defense economics. Specifically, this report addresses the prospects for economic growth, military spending, and military investments in Germany, France, the United Kingdom, and Italy to provide sufficient resources for enhancing the military capabilities sought by these four principal European NATO members for the ESDP/RRF in the EU, or for the DCI within NATO.

Our analysis of economic and military trends in these four countries begins by focusing on the record of the 1985–1999 period and uses this analysis as a basis for corresponding forecasts for the ensuing decade, from 2001 to 2010. The analysis of the past 15 years, as well as our forecasts for the forthcoming decade, employ an aggregate model in which gross domestic product (GDP) is estimated from an aggregate production function for each country; military spending is estimated as a specified, sometimes varying, proportion of GDP; and military investment is estimated as a specified, sometimes varying, proportion of military spending.

The data used in exercising the model have been collected from multiple national and international sources that not infrequently disagree with one another. As suggested above, our analysis of the historical trends is used to generate several key parameters which, with adjustments that we explain, are employed to forecast GDP, military spending, and military investments in the forthcoming decade. These estimates in turn are compared with independently derived estimates of the investment costs of the ESDP/RRF force to assess prospects in the four countries for meeting the incremental investment costs of the targeted force. We formulate several options that would enable these countries, in conjunction with other non-NATO members of the EU, to finance the investment costs associated with the ESDP/RRF.

The method we have followed in EMPEC is summarized in the schematic on the following page.

Analysis of the trends in economic growth and military spending in the four countries over the 1985–1999 period highlights several general points that cover the 15-year trend:

- GDP growth fluctuated between low and moderate levels during the period, accompanied by high volatility as reflected by large standard deviations in all four countries.
- Economic growth was accompanied by very low growth in aggregate factor productivity, reflecting relatively slow rates of technological progress, yet moderately high growth of labor productivity.

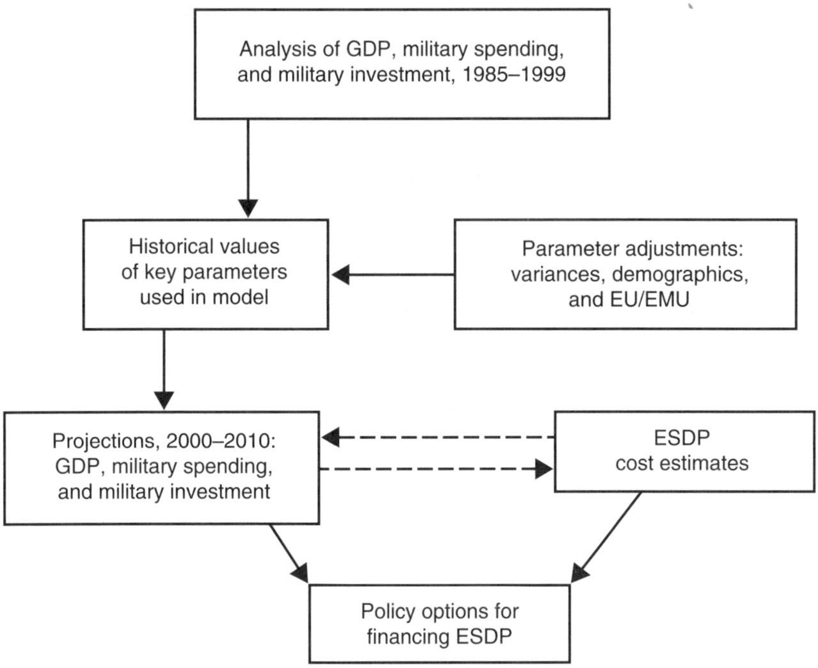

- These characteristics reflected, in turn, minimal growth of employment, together with high rates of growth in capital formation and hence increases in capital-labor intensities.
- Finally, military spending declined monotonically as a share of GDP in all four countries and in terms of real spending levels, except in Italy, whose real military spending increased by just under 1 percent per year.

Our forecasts of economic growth envisage annual average growth rates over the next decade of 2.4 percent for Germany, 2.25 percent for France, 2.32 percent for the United Kingdom, and 1.62 percent for Italy. We forecast that military spending during this period could rise modestly as a consequence of the additional resources generated by economic growth, and that military investments (i.e., procurement plus research, development, testing, and evaluation, RDT&E) in the four countries considered together might rise from their aggregate 2000 total of about $28 billion by an additional $5.3 billion in the years from 2001 to 2003, an additional $17.6 billion between 2004 and 2007, and an additional $21.6 billion between 2008 and 2010 (all expressed in U.S. 2000 dollars).

These forecasts thus depart from the consensus view of many of our European interlocutors that real levels of military spending and military investment will remain fixed at their present levels or will decline, rather than increasing slightly as a consequence of economic growth.

There are formidable obstacles to estimating the costs of the planned ESDP/RRF. First, no cost estimates have yet been made by the EU. Second, while many meetings and conferences have been held to address the purposes, design, and general capabilities that the force is intended to have, and several communiqués have been issued, the capabilities in question have been described in verbal rather than quantitative terms. For example, the enhanced capabilities have been described in terms of expanded airlift, improved C^4I capabilities, rapid deployability, advanced air and surface missile defense, and interoperability with U.S. forces, but there has been nothing approaching tables of organization and equipment (TO&E) for the new forces.

In attempting to make some preliminary cost estimates in light of these obstacles, we have focused solely on the associated costs of military investments—that is, military procurement and RDT&E. We have not considered operations and maintenance (O&M) costs, which might entail an additional 30 to 50 percent of the capital costs associated with equipping the enhanced ESDP force, or alternatively might replace O&M costs of existing military forces. To arrive at reasonable conjectures concerning the incremental investment costs and prospective burdens on European defense budgets, we have applied four approaches:

- A "bottom-up" approach focusing on the major systems and acquisitions that the force is expected to require, identified from discussions within the EU.
- A generalized "top-down" approach using the procurement and RDT&E costs per member of the U.S. armed forces as a basis for estimating the corresponding costs of the 60,000-man ESDP/RRF.

- An approach using a rough estimate of the capital costs associated with a U.S. Marine brigade as a plausible building block for the ESDP force.
- An approach similar to the preceding one, but using a RAND costing model with a U.S. Army assault division as the modular building block.

The several preliminary and crude approaches to estimating the capital costs of the ESDP force produce a range of $24 billion to about $56 billion (in U.S. 2000 dollars).

Consideration is then given to potential sources of funding to meet these costs, focusing on three prominent sources: (1) incremental resources that may be generated by military investments resulting from economic growth; (2) resources that may be drawn from reallocation of existing military investment budgets; and (3) savings that may be realized by consolidation of the EU defense industry and by a unified European defense market.

The following conclusions emerge from this examination of potential funding sources:

1. Meeting the capital costs of the ESDP/RRF by the planned 2003 target date is unlikely.
2. Assuming that incremental investment resources are available, but without substantial reallocations from existing military spending and military investments, the requisite capital costs for the enhanced force cannot be met until the end of the 2001–2010 decade.
3. With such reallocations, the necessary capital costs can be met by 2007.
4. With combinations among the several funding sources, including achievement of a common defense market and consolidation of the European defense industry, reaching the ESDP/RRF goals could be further accelerated.

To move aggressively in this direction will require overcoming organizational inertia and established service and industry resistance. The political will to accomplish this may be forthcoming. However, thus far, the rhetoric behind the ESDP has proceeded far more rapidly than has the acquisition of the resources required to turn the concept into a reality, whether through the provision of additional resources or the reallocation of existing resources.

In sum, we conclude that the United States and NATO probably have less reason to worry that the EU will acquire genuinely enhanced capabilities for the ESDP force than that it will continue to produce descriptive rhetoric without the resources necessary to acquire and support them.

ACRONYMS

AAR	after-action review
AGS	armored gun system
C^4I	command, control, communications, computers, and intelligence
CINC	Commander in Chief
CSAR	combat search and rescue
DCI	Defense Capabilities Initiative
ELINT/SIGINT	electronic intelligence/signals intelligence
EMPEC	European military prospects and economic constraints
EMU	Economic and Monetary Union
ESDP	European Security and Defense Policy
EU	European Union
EW	electronic warfare
GDP	gross domestic product
GNP	gross national product
GPF	ground processing facility
ILO	International Labor Organization

IMF	International Monetary Fund
MEB	Marine expeditionary brigade
MEF	Marine expeditionary force
O&M	operations and maintenance
RDT&E	research, development, testing, and evaluation
RRF	Rapid Reaction Force
SACEUR	Supreme Allied Commander Europe
SEAD	suppression of enemy air defenses
TO&E	tables of organization and equipment
UAV	unmanned aerial vehicle

Chapter One

BACKGROUND AND OBJECTIVES

RAND's study of European military prospects and economic constraints (EMPEC) was motivated by the concern of U.S. policymakers with moving toward a better balance with our European allies in providing military capabilities, as well as in the sharing of responsibilities, decisionmaking, and costs of our alliance relationships. This concern is largely shared by our NATO allies, who also nurture—with varying degrees of intensity—the additional aim of enhancing the collective military capabilities of the European Union (EU) apart from the NATO alliance, while at the same time moving toward a better balance within the alliance itself.

The principal focus of the EMPEC project is on the EU's European Security and Defense Policy (ESDP) and the ESDP's instrument, the Rapid Reaction Force (RRF), with secondary attention to the Defense Capability Initiative (DCI) within NATO and the relationship between the ESDP and the DCI. The DCI is a set of commitments by the non-U.S. members of NATO to upgrade their individual and collective capabilities and thereby enhance their capacity for bearing a larger share of the burden of and responsibility for NATO operations.

Although these issues have many important political, diplomatic, and legal aspects, the narrower focus of this study is on defense economics. More specifically, we address the prospects for economic growth, military spending, and military investments in Germany, France, the United Kingdom, and Italy to provide sufficient resources to finance the enhanced military capabilities sought by these four principal European NATO members—whether the capabilities are attributed to the ESDP/RRF in the EU, or to the DCI within NATO.

The ESDP within the EU is a complex mixture of rhetoric, strategy, politics, and, to a lesser extent, economics. A typical example of the *rhetoric* associated with the ESDP is the following pronouncement in the EU's Military Capabilities Commitment communiqué of November 2000:

> [A] priority of the Union [is] to develop and introduce the civil and military resources and capabilities required to enable the Union to take and implement decisions on the full range of conflict-prevention and crisis-management missions.[1]

In the realm of ESDP *strategy*, the same source asserts that:

> these forces should be militarily self-sustaining with the necessary command, control and intelligence capabilities, logistics, other combat support units, and, as required, air and naval elements.[2]

The complexities and refinements of the *politics* of the ESDP (as well as its not always transparent relationship to the DCI and NATO) were deftly formulated by British Defense Secretary Geoffrey Hoon:

> [W]e have been negotiating with our allies to ensure that any proposals for a rapid reaction capability inside the European Union are completely consistent with the NATO planning process.[3]

U.S. Secretary of State Colin Powell expressed the political stance of the United States toward the ESDP emphatically, if redundantly:

> We will support any such effort [to improve European defense capabilities] as long as it strengthens NATO, and does not weaken NATO.[4]

---

[1]Military Capabilities Commitment Conference, *Conference Communiqué*, Brussels, November 2000.

[2]Ibid.

[3]Statement in the British House of Commons, December 2000.

[4]Senate Confirmation Hearing, *U.S. Senate*, January 2001.

The evident, but not dominant, economic dimensions of the ESDP, and implicitly also of the DCI, were suggested by former U.S. Secretary of Defense William Cohen:

> [I]f our European allies are to close the distance with American [defense] technology, they simply must make a greater investment in national security by reallocating scarce resources, committing to regular upgrading of equipment, and increasing funding of research and development. At the very least, budgets must be restructured to generate funds for new spending.[5]

Javier Solana, the EU's High Representative for Security Affairs and a former Secretary-General of NATO, echoed this sentiment:

> It is important that we explain domestically why supplementary efforts are needed [and military budgets have to be increased].[6]

Through NATO and the DCI, and through the ESDP, European members of NATO, as well as the non-NATO members of the EU, have expressed their intention and commitment to enhance Europe's military capabilities. The result of their doing so could be a more balanced sharing of security burdens and responsibilities within NATO and, if circumstances warrant, a capacity for independent action by the EU.

Against this background, the EMPEC project focuses on the defense economics of the ESDP and also, indirectly, of the DCI. While the two initiatives differ, they share a requirement for additional military capabilities and hence for additional military investments. They are thus separately and collectively part of the defense economics of the four largest European countries. This focus comprises the bulk of our report, although we conclude with observations concerning some noneconomic dimensions as well. In combination, all of these aspects will significantly influence the character of U.S.–European security relations during the next decade.

---

[5]William Cohen, speech before the International Institute for Strategic Studies, San Diego, California, September 1999.

[6]Javier Solana, statement made at the Brussels Commitment Conference, November 2000.

We begin in Chapter Two by examining the economies of the four countries and summarizing the method and model we have used to analyze economic trends of gross domestic product (GDP) growth, military expenditures, and military investments from 1985 to 1999. In turn, this analysis provides the basis for our forecasts of the three corresponding variables—GDP, military spending, and military investments—in Germany, France, the United Kingdom, and Italy in the period from 2001 to 2010.

Chapter Three summarizes our results for the 1985–1999 period, the several key parameters drawn from this historical analysis, and the adjustments we have made in them to provide the values of these parameters that we use in our forecasts of GDP, military spending, and military investment in the ensuing decade.

Chapter Four then presents several different—but all problematic—estimates of the capital costs associated with the planned ESDP/RRF/DCI enhancements, together with a formulation and brief evaluation of various possible options for providing EU funding to meet these costs.

Chapter Five concludes with several observations about the political and strategic issues connected with the planned enhancement of European military capabilities, as well as the implications and conclusions about these issues that may be inferred from our analysis of the defense economics that bears on them.

Chapter Two

EUROPEAN ECONOMIES AND DEFENSE BUDGETS, 1985–1999 AND BEYOND: METHODS AND MODELS

Our analysis of economic and military trends in the four countries over the 1985–1999 period, as well as our forecasts for the ensuing decade, are based on an economywide model in which (1) GDP is estimated from an aggregate production function for each economy, (2) military spending is estimated as a specified, sometimes varying, proportion of GDP, and (3) military capital is estimated as a specified, sometimes varying, proportion of military spending that is allocated to new military investment.

This model was selected for its commendable transparency, its convenience for repetitive calculation purposes, and its relatively modest and tractable data requirements compared, for example, with more elaborate multisectoral, input-output, and other disaggregated models. Specification of the model and other methodological details are summarized in Appendix A.

For the 1985–1999 period, we have collected from multiple sources time-series data for Germany, France, the United Kingdom, and Italy covering their GDPs, demographics, employment, capital formation, defense budgets, and military investment. The several sources, the data from which do not always coincide, include the International Monetary Fund (IMF), the Organization for European Cooperation and Development (OECD), the International Labor Organization (ILO), each country's statistical yearbooks, NATO, and the U.S. Census Bureau. The data are inserted in the model described in Appendix A to generate seven key parameters and their variances in the 1985–1999 period: annual GDP growth, annual employment growth,

growth in the capital stock, wage (and nonwage) distributive shares, rate of growth in total factor productivity, military spending shares of GDP, and military investment shares of military spending. In turn, these parameters provide the inputs we use in making forecasts for 2001–2010.

The forecasts of GDP, military spending, and military investment for the 2001–2010 period use the historically calculated parameter values, which we have adjusted in some instances for reasons described below. The parameter adjustments have been influenced by three factors: the variances or volatility of the historical parameters as originally estimated from the time-series data referred to above; the demographics, changing composition of the labor force, and other specific conditions in each country; and our assessment of prospective effects of the EU and the Economic and Monetary Union (EMU) on the competitiveness and productivity of labor and capital in each country.

The method we have followed is depicted schematically in Figure 1.

Analysis of the historical trends of GDP, military expenditures, and military investments covered in Chapter Three is shown in the top tier of Figure 1. Based on the model described in Appendix A, the historical values of the key parameters listed above are represented in the second tier. Also in the second tier of the figure, we refer to the adjustments made in these parameters based on their variances, demographics, and the effects of the EMU, reflected by the horizontal arrow. In turn, these adjusted parameters are used in the model to generate the forecasts presented in Chapter Three of GDP growth, military spending, and annual military investment in the forthcoming decade, as shown in the third tier of Figure 1. Finally, the box and dotted interactive arrows in the figure refer to the several approaches we have used in estimating the costs of the ESDP (addressed in Chapter Four) and their comparison with the forecasts. This comparison provides the "bottom line" of the report, which includes the formulation of several options for financing ESDP costs and the corresponding time periods required to accomplish this (shown in the fourth tier).

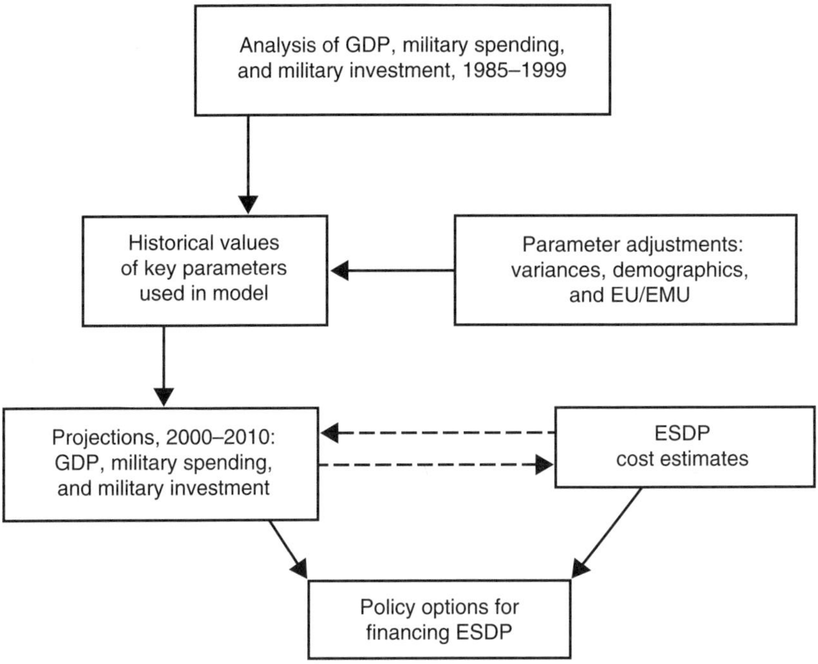

Figure 1—Trends, Forecasts, and Policy Options

Chapter Three

ECONOMIC AND MILITARY TREND ANALYSIS, 1985–1999, AND FORECASTS FOR 2001–2010

EUROPEAN ECONOMIC GROWTH AND MILITARY SPENDING, 1985–1999

Our analysis of the economic and military trends in the four countries over the past 15 years has focused on seven parameters which provide the basis for the forecasts we make for the forthcoming decade. These are the annual averages and standard deviations over the 1985–1999 period of GDP growth, employment growth, capital growth, growth in total factor productivity (i.e., the combined productivity of the weighted labor and capital inputs), the labor share in gross income, the military spending share of GDP, and the military investment share of military spending. In turn, adjusted values of six of these parameters (excluding GDP) provide the basis for the forecasts summarized in this chapter.

Table 1 shows these parameter values for the 1985–1999 period. Table 2 shows military spending levels (in constant U.S. 2000 dollars) for the four countries for 1985, 1990, 1995, and 1999.

The 15-year period for which these trends have been calculated reflects several unusual characteristics.[1] For example, the Cold War confrontation between the West (NATO) and the East (the Soviet

[1]We are indebted to Lord John Roper for emphasizing the abnormal features of the 1985–1999 period, although the answer to the question of what precisely would constitute normality is elusive.

Table 1

European Economic Trends and Parameter Values, 1985–1999: Germany, France, United Kingdom, and Italy

	GDP Growth (percent/year)		Employment Growth (percent/year)		Capital Growth (percent/year)		Total Factor Productivity Growth (percent/year)		Labor Share (percent)	Military Spending Share of GDP (percent/year)		Military Investment Share of Military Spending (percent/year)	
	Mean	SD	Mean	SD	Mean	SD	Mean	SD	Mean	Mean	SD	Mean	SD
Germany (1991–1999)[a]	2.62	4.19	–0.72	0.6	3.0	0.9	0.17	1.01	65	1.83	0.26	18.6	0.80
France	2.18	1.42	0.3	0.7	5.1	0.4	0.46	(b)	70	3.5	0.4	30.0	(c)
United Kingdom	2.67	2.01	0.9	1.9	4.3	0.5	0.42	(b)	65	3.5[d]	0.9	30.0	2.4
Italy	1.92	1.37	–0.1	1.6	1.5	1.0	0.18	1.01	55[e]	2.1	0.1	22.1	2.9

SOURCES:

GDP: International Monetary Fund, *International Financial Statistics Yearbook 2000*, 1999.

Exchange rates: International Monetary Fund, *International Financial Statistics Yearbook 2000*, 1999.

Capital formation and capital growth estimates based on assumption that capital stock in 1985 and subsequent years is four times GDP *less* annual depreciation of 8 percent and *plus* annual capital formation, *or* six times capital formation in 1969 (see discussion in Appendix A), International Monetary Fund, *International Financial Statistics Yearbook 2000*, 1999.

GDP deflator: International Monetary Fund, *International Financial Statistics Yearbook 2000*, 1999; U.S. Council of Economic Advisors, *Economic Indicators*, various issues; *Annual Report of the Council of Economic Advisors*, January 2001.

Population, labor force, employment: OECD, *Labor Force Statistics,* 1998; ILO, *Yearbook of Labor Statistics,* 1999; U.S. Bureau of the Census, *International Database,* 2000.

Labor shares in GDP:

Germany–OECD, National Accounts Database, OECD, *National Accounts*, Vol. II, 1998; discussions with Mr. Stephan Hauf, *Statistik–Bund;* authors' econometric estimates.

France–Authors' estimates based on French National Accounts and OECD.

United Kingdom–Office for National Statistics, United Kingdom National Accounts, *The Blue Book,* 1999.

Italy–OECD, National Accounts Database, OECD, *National Accounts,* Vol. II, 1998; Bank of Italy, *Annual Report,* 1999, Statistical Appendix; authors' econometric estimates.

Historical growth rates for total factor productivity: authors' econometric estimates.

Military expenditures, military investment: NATO Press Release, December 2, 1999; financial and economic data relating to NATO defense, *Atlantic News,* No. 3254 (Annex), December 13, 2000; U.S. Arms Control and Disarmament Agency/Bureau of Verification and Compliance, *World Military Expenditures and Arms Transfers,* various issues; International Institute for Strategic Studies, *The Military Balance,* various issues; U.S. Office of Management and Budget, *The Budget for Fiscal Year 2002,* Historical Tables; Joseph E. Nation, *German, British, and French Military Requirements and Resources in the Year 2005,* Rand Graduate School Dissertation, M–2982–RGSD, Santa Monica, California, 1992.

[a]Data from 1969–1990 are for FRG only, excluding East Germany; data for 1991–1999 reflect German reunification.

[b]Total factor productivity was computed for the 15-year period as a whole, rather than year-to-year, based on the mean values of employment growth, capital growth, and GDP growth.

[c]France's military investment share was assumed to be the same as that of the United Kingdom, rather than computed, because of noncomparabilities in the French data.

[d]United Kingdom military spending share of GDP decreased annually from 1985 to 1999, except in 1998, when it remained the same as in 1997.

[e]Approximate midpoint between the Bank of Italy (65 percent) and the OECD (42 percent).

Table 2

European Military Spending, 1985–1999: Germany, France, United Kingdom, and Italy (billions of U.S. 2000 dollars)

Military Spending	Germany	France	United Kingdom	Italy
1985	NA	39.4	44.8	18.2
1990	38.1 (1991)	40.1	37.8	20.5
1995	31.2	37.2	29.3	18.1
1999	32.7	37.4	27.0	22.2
Ratio, 1999/1990 (percent)	86	93	71	108
Average annual change, 1990–1999 (percent/year)	–1.7	–0.7	–1.7	0.9

SOURCES: See Table 1.

Union and the Warsaw Pact) dominated the first five years of the period. Unification of West and East Germany, dating from 1991, and its huge costs imposed another abnormal burden on Germany during the period. Because of these unusual circumstances, one should be cautious in using and adjusting parameters derived from the 1985–1999 period to make forecasts for the forthcoming decade. That having been said, it is also worth noting that what might be considered "unusual" circumstances frequently characterize the international environment. We have tried to exercise caution in adjusting the trend parameters summarized later in this chapter.

Nevertheless, granting the unusual features of the trend period, several salient points are worth noting about the data summarized in Tables 1 and 2 for each of the four countries.

Germany

- GDP growth in the unification period from 1991 to 1999 was moderate (about 2.62 percent annually, with a standard deviation about three times this magnitude; in 2000, Germany's GDP growth was 3.0 percent;[2] exclusion of 1991 reduces these figures to 1.3 percent and 1.5 percent, respectively).

- Average employment during the 1990s fell by a bit less than 1 percent per year, with a standard deviation of 1 percent.
- Capital growth was moderately high, 3 percent annually, with volatility about one-third that amount; growth of the German capital stock was lower in the late 1990s than in the early part of the decade.
- Factor productivity grew at a small fraction of 1 percent during the 1990s, although average labor productivity grew at an annual rate of about 2 percent, with a standard deviation of 1 percent.
- Military spending declined monotonically as a percentage of GDP, from 2.8 percent at the start of the 1990s to 1.6 percent at the end of the decade. In real terms, annual military spending declined by about 15 percent during this period. The share of military investment in total military spending was about 18 percent during the 1990s.

France

- Annual GDP growth in France over the 1985–1999 period averaged just over 2 percent (although in 2000, France's growth rate was over 3 percent), with a standard deviation somewhat lower than the average annual growth rate.
- Employment growth during the 15-year period was about one-third of 1 percent annually, accompanied by a high unemployment rate, averaging between 9 and 10 percent of the labor force throughout the period.
- France's capital stock grew at a high rate, over 5 percent per year, with a standard deviation less than one-tenth of this rate.
- Total factor productivity growth was slightly below one-half of 1 percent annually during the trend period, although labor pro-

[2]As indicated in the notes to Tables 1 and 2, the data used in our analysis of the German economy are confined to the post-unification period, 1991–1999, while the data used for the other three countries cover the entire 1985–1999 period.

ductivity grew at an average annual rate of 1.6 percent due to capital deepening (i.e., increased capital-to-labor ratios).

- Military spending as a share of GDP decreased monotonically over the 1985–1999 period, from 4 percent in 1985 to slightly less than 2.8 percent in 1999, representing a decrease in real military spending of about 7 percent during this period.

United Kingdom

- GDP grew at a moderately high average annual rate, 2.67 percent, but with volatility nearly equal to the average growth rate. In 2000, GDP growth was 3 percent.
- Employment growth was just under 1 percent annually over the 15-year period; this was high relative to that of other EU countries, but volatility was also high—about twice the rate of employment growth.
- Growth of the United Kingdom's capital stock was over 4 percent per year, with very low volatility (standard deviation equal to 0.5 percent).
- Total factor productivity grew at a rate of four-tenths of 1 percent per year, although average labor productivity grew by nearly 2 percent annually, again due to capital deepening.
- Military spending as a share of GDP decreased from over 5 percent in 1985 to just above 2 percent in 1999, with real military spending decreasing by 40 percent from 1985 to 1999.

Italy

- Italy's GDP growth averaged slightly below 2 percent during the trend period, with markedly slower growth in most of the 1990s compared with the previous (1985–1990) period. In 2000, Italy's GDP grew by 2.8 percent.
- Annual employment growth was slightly negative at –0.13 percent over the period, accompanied by high volatility.
- Capital growth was 1.4 to 1.5 percent annually, a low rate relative to that of the other three countries.

- Total factor productivity growth was also very low at 0.18 percent per year, although average labor productivity grew at an annual rate of more than 2 percent.
- Military spending showed a moderate decline as a share of GDP during the 1990s. The military investment share of military spending exhibited a monotonic decline, from over 25 percent in 1985 to slightly over 18 percent in 1999.

Several generalizations can be made about the specific points relating to each of the four countries during the 15-year period:

- In all four countries, GDP growth fluctuated between low and moderate levels, accompanied by moderate-to-high volatility, as reflected by high standard deviations.
- Economic growth was accompanied by very low growth of total factor productivity, reflecting relatively slow rates of technological progress but moderately high growth of labor productivity.
- These characteristics reflected, in turn, minimal growth of employment, together with high rates of growth in capital formation, increases in capital-to-labor ratios, and inefficient capital investments.
- Finally, military spending declined monotonically, both as a share of GDP in all four countries over the 1985–1999 period and in terms of real spending levels, except in the case of Italy, whose real military spending increased slightly.

FORECASTS OF GDP AND MILITARY SPENDING, 2001–2010

Our forecasts of GDP, military spending, and military investment for the period 2001–2010 are based on the historical parameter values summarized in Table 1, with small adjustments that we believe are prudent in these parameters.

As noted in Table 1 and Table 3 below, these adjustments have been based on three considerations: First, where volatility of the historical parameters was high, as reflected by their standard deviations, we lowered the parameters slightly. Second, adverse changes in each country's demographics, and in particular the impending increases in dependency ratios,[3] resulted in our lowering the expected rates of growth in total factor productivity, employment, and capital formation. Finally, our assessment of the effects of the EMU on the prospective efficiency or inefficiency of resource allocations in the EU is also reflected in small reductions to be expected in total factor productivity growth in the next decade.

The demographic data affecting the parameter adjustments used in our forecasts are shown in Appendix B.

DEMOGRAPHIC EFFECTS

As the charts in Appendix B indicate, each of the four countries will encounter rising dependency ratios in the composition of their populations. Between 2001 and 2010, dependency ratios in Germany will rise from approximately 42 percent to 45 percent; in France, from 38 percent to over 43 percent; in the United Kingdom, from 38 percent to 42 percent; and in Italy, from 41 percent to 49 percent. The resulting increase in the proportion of retirees to employed labor, as well as the expected burden of social entitlements and its implications for the level and composition of taxation, will affect adversely several of the parameters' historical values. Higher tax rates are likely to have similar, if small, negative effects on capital formation and on rates of growth of factor productivity.

Consequently, in light of these considerations, we have made the following changes in the parameter values derived from the analysis of the 1985–1999 period:

- The employment growth rates are adjusted slightly downward.
- The wage-share parameters are adjusted slightly upward (reflecting the declining proportion of the population in the prime working-age cohort, leading to increases in expected wage rates).
- Rates of growth in the capital stock and in total factor productivity are adjusted slightly downward.

---

[3]Dependency ratios represent the ratio between the population over 60 years of age and the 20- to 59-year age group. See Appendix B.

EMU/EU EFFECTS

The final set of considerations affecting adjustments in the historical parameter values relates to the effects of the EMU on the outlook for EU productivity and competitiveness. Assessment of these effects is complicated because there are likely to be both positive and negative effects. Deriving a net assessment of these effects is uncertain and controversial.

The benefits to be expected from the single Euro currency include avoidance of transaction costs associated with the multiple currencies of the EMU members, elimination of future competitive devaluations to promote exports by members of the EMU, and elimination of the risk of intra-EMU exchange-rate adjustments and hence of the costs of hedging against such risks.

There may also be significant negative effects from the monetary union. Use of the single currency means that the heterogeneous and structurally dissimilar economies of EMU members will be impeded in their respective abilities to adjust to such external shocks as changes in world oil prices and global deflation or inflation. Increased disparities in economic growth and unemployment rates among the members are likely to ensue, perhaps generating political pressures for inflationary monetary expansion. Continued rigidities in European labor markets add a further impediment to smooth adjustments to such shocks through wage variation or labor migration within the EU.

The resulting increase in diversity of economic conditions within the EU is likely to generate increased political pressures for fiscal redistribution to alleviate these differences. In turn, these and other pressures may tend to increase the central power of the European Commission,[4] probably boosting the required contributions by member nations and perhaps leading to direct taxation imposed by the EU itself. If the tax burden were to increase as a result, the effect would be to lower long-term economic growth. A further consequence of the EMU is likely to be an increase in trade within the EU at the expense of trade with non-EMU countries, which could create

---

[4]German Chancellor Gerhard Schroder's recent proposal for reforming the EU would decidedly move in this direction. See *New York Times*, May 1, 2001.

additional trade distortions and inefficiencies in resource allocations among the member economies.

Netting out these prospective benefits and losses from the single currency is difficult. Our assessment, which marginally affects the parameter adjustments, is that the single currency will be likely to

- Slightly lower the rates of employment growth.
- Similarly lower rates of growth in the capital stock (because a higher tax burden is likely to be realized in part at the expense of corporate profits and capital formation).
- Finally, slightly lower the growth of total factor productivity.

Table 3 and Figures 2 through 5 summarize our forecasts for GDP, military expenditures, and military investment in Germany, France, the United Kingdom, and Italy for the period from 2001 to 2010, expressed in constant U.S. 2000 dollars. These forecasts reflect the previously discussed considerations and are further explained in the footnotes to Table 3.

Table 3 also highlights the *incremental* military investments above the estimated military investments (i.e., military procurement plus RDT&E) in the year 2000 in each of the four countries. We emphasize incremental military investment above estimated investment in 2000 in order to calibrate the additional resources that would be generated by economic growth. In principle, these resources could contribute to meeting the capital costs of the ESDP/RRF, as well as commitments already made within NATO under the DCI, if the share of GDP allocated to military spending remained constant while little or no reallocation within the existing patterns of military investments was actually realized. To the extent that appreciable reallocations from existing and planned military investments in each of the countries can be accomplished, the problem of providing resources to finance high-technology equipment for the ESDP force would be substantially eased. Thus, the incremental military investments shown in Table 3 imply a "worst-case" scenario where inertia predominates, pressures from established service and industry interests to maintain their shares of the military "pie" prevail, and existing patterns of procurement outlays for tanks, artillery, missiles, aircraft, and logistic support remain relatively unperturbed.

Table 3

Current and Outyear Estimates, 2001–2010 (billions of U.S. 2000 dollars, except as indicated)

	Germany	France	United Kingdom	Italy	Total
GDP average growth rate (percent per year)	2.4	2.25	2.32	1.62	
GDP					
2001	1,868	1,396	1,217	1,130	5,611
2005	2,060	1,524	1,333	1,211	6,128
2010	2,319	1,704	1,493	1,294	6,810
Military spending					
2001	34.2	30.1	28.0	23.2	120.1
2005	37.7	32.9	30.7	25.1	126.4
2010	42.4	36.6	34.3	27.1	140.4
Military investment, 2000	6.8	8.8	8.2	4.2	28.0
Incremental military investments (above 2000 level)					
2001–2003	1.0	1.22	1.18	1.3	4.7
2004–2007	3.9	4.62	4.42	4.7	17.64
2008–2010	4.9	5.83	5.63	5.0	21.63

SOURCES: See Table 1. Forecasts are based on the model described in Appendix A. All conversions from own currencies are in constant 2000 prices, using 2000 nominal exchange rates. (See the discussion of own-currencies' conversions to U.S. 2000 dollars in Appendix A.) The key parameter values used in the forecasts are as follows (all figures are in percentages):

	Germany	France	United Kingdom	Italy
Employment growth ($\dot{L}/L$)	0.05	0.3	0.8	0.13–0.19 (0.15 avg.)
Capital growth ($\dot{K}/K$)	2.5–3.02	5.0	4.0	2.4
Total factor productivity (τ)	0.175–0.20[a]	0.46	0.4	0.2
Labor share (α)	65	70	70	55
Military spending share (δ)	1.83	2.2	2.3	2.09
Military investment share (π)	20.3	30[b]	30[b]	20.7

[a]The growth rate of German total factor productivity is assumed to increase gradually over 2001–2010 from 0.175 to 0.20 percent at a constant labor share of 65 percent.

[b]The military investment shares of military spending for the United Kingdom and France have been raised from their estimated 1985–1999 levels to allow for additional RDT&E connected with acquisition of technologically more-advanced systems.

As Table 3 indicates, our forecasts for average annual GDP growth rates for the 2001–2010 period are: 2.4 percent for Germany (compared to 1.8 percent during the 1991–1999 period), 2.25 percent for France (compared to the historical 1985–1999 average of 2.0 percent), 2.3 percent for the United Kingdom (compared to the actual 1985–1999 figure of 2.5 percent), and 1.62 percent for Italy (compared to an actual figure of 2.0 percent for the 1985–1999 period).

Figures 2 through 5 show the trajectories of our forecasts of GDP, military expenditures, and military investments in each year of the forthcoming decade for the four countries.

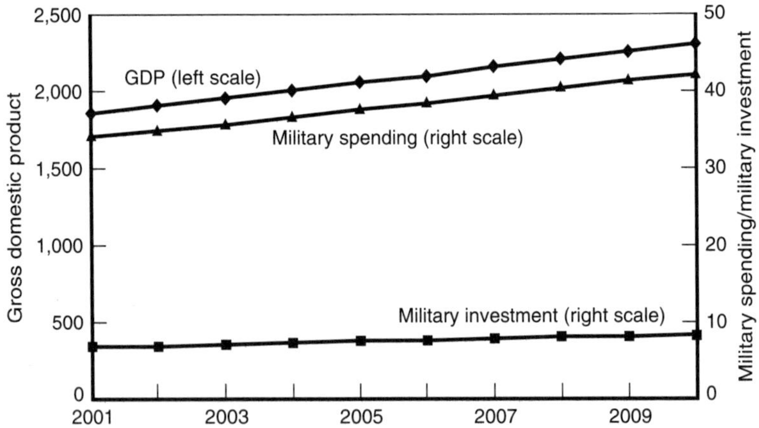

NOTE: Conversions from German currency in constant 1995 prices, using 1995 nominal exchange rates, then applying U.S. GDP deflator to express figures in U.S. 2000 dollars. The smooth trajectories of the series are a consequence of our omission of cyclical or other perturbations that would generate ups and downs in the actual year-to-year results.

Figure 2—Germany: GDP, Military Spending, and Military Investment, 2001–2010 (billions of U.S. 2000 dollars)

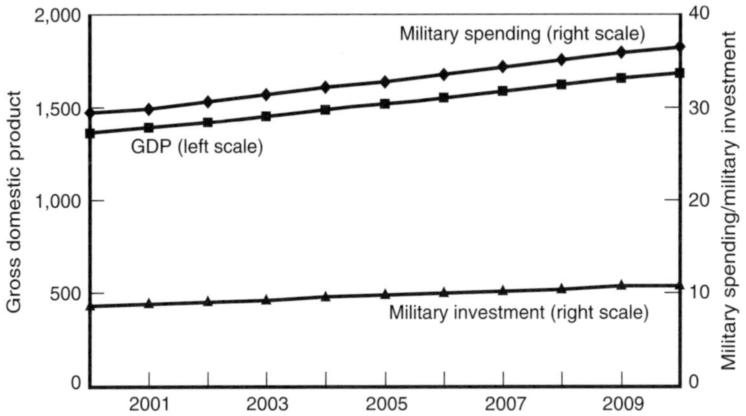

NOTE: Conversions from French currency in constant 1995 prices, using 1995 nominal exchange rates, then applying U.S. GDP deflator to express figures in U.S. 2000 dollars. The smooth trajectories of the series are a consequence of our omission of cyclical or other perturbations that would generate ups and downs in the actual year-to-year results.

Figure 3—France: GDP, Military Spending, and Military Investment, 2001–2010 (billions of U.S. 2000 dollars)

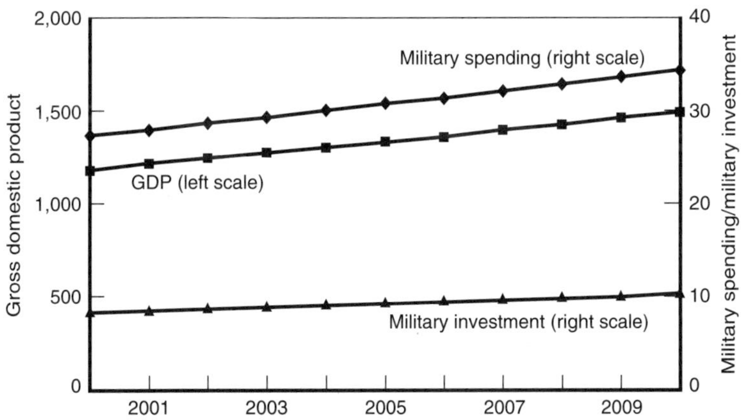

NOTE: Conversions from U.K. currency in constant 1995 prices, using 1995 nominal exchange rates, then applying U.S. GDP deflator to express figures in U.S. 2000 dollars. The smooth trajectories of the series are a consequence of our omission of cyclical or other perturbations that would generate ups and downs in the actual year-to-year results.

Figure 4—United Kingdom: GDP, Military Spending, and Military Investment, 2001–2010 (billions of U.S. 2000 dollars)

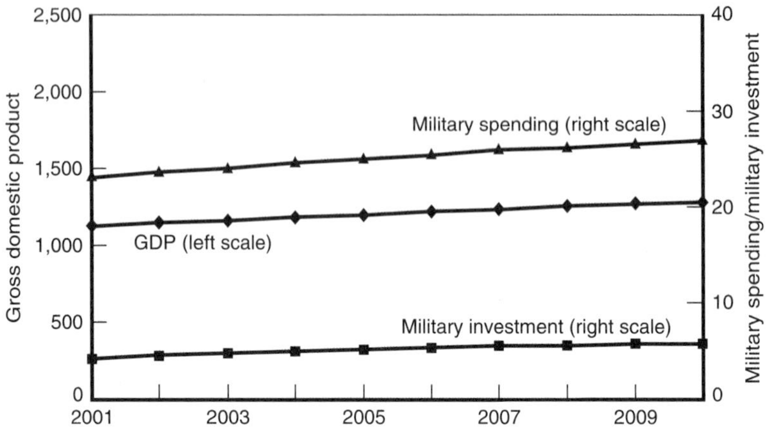

NOTE: Conversions from Italian currency in constant 1995 prices, using 1995 nominal exchange rates, then applying U.S. GDP deflator to express figures in U.S. 2000 dollars. The smooth trajectories of the series are a consequence of our omission of cyclical or other perturbations that would generate ups and downs in the actual year-to-year results.

Figure 5—Italy: GDP, Military Spending, and Military Investment, 2001–2010 (billions of U.S. 2000 dollars)

These forecasts indicate that the *incremental* military investment that would plausibly be available for meeting the capital costs of the ESDP force—above *actual* military investment outlays in 2000—amount to approximately $5.3 billion in the 2001–2003 period, an additional $17.6 billion accumulating during the 2004–2007 period, and an additional $21.6 billion accumulating in the 2008–2010 period. If appreciable changes can be made in the current pattern of allocations for military investments, there is a pool of approximately $28 billion in year 2000 annual military investments in the four countries that could be tapped (say, by one-third each year) to provide funding for the enhanced ESDP capabilities, rather than having to rely only on *incremental* investment for ESDP financing.

It should be noted that the assumptions embedded in our forecasts diverge in two significant respects from views that are widely held among defense policy experts in the four countries. First, based on our discussions, their consensus is that *real levels of military spend-*

ing and military investment will be fixed at their present levels or reduced, rather than rising slightly, as we have assumed, as a consequence of aggregate economic growth and a constant share of GDP allocated to defense purposes.

For example, our German interlocutors emphasized that paying down Germany's public debt—swollen by the burden of financing reunification costs—is a higher-priority claimant for budgetary resources than is any addition to military spending.

Second, the prevailing European view appears to be that there are *realistic opportunities for reallocation within existing patterns of military spending and military investments, thereby allowing for shifting resources from the existing pool of military procurement without a need for additional resources.*

In contrast to these views, we have assumed that reallocations from existing patterns to meet the differing procurement and RDT&E costs associated with the ESDP/RRF/DCI are likely to be quite limited, because of organizational inertia, powerful service interests, and the familiar "iron triangle"—no less familiar in Europe than in the United States—between the defense industry, the military services, and national legislatures. On the basis of this premise, we have therefore emphasized the need for additional resources to meet the capital costs of the planned force through moderate additions to military spending and military investment budgets.

Chapter Four

PRELIMINARY COST ESTIMATES OF THE ESDP/RRF AND POTENTIAL SOURCES OF FUNDING

When considering reasonable cost estimates for the ESDP/RRF, three observations are paramount. First, no cost estimates have been made by the EU, although discussions within the EU of the proposed force have been going on for several years. The EU force has been discussed at innumerable meetings of the European Council and at various conferences, committee meetings, and working group sessions, and numerous communiqués have been issued. These have dealt with the purposes, design, and general capabilities that the force is intended to have, the scale of manpower for the force (between 50,000 and 60,000), and pledges of support and commitments for it, in conformity with the Petersberg tasks established in 1992. However, specific tables of organization and equipment (TO&E) and cost estimates associated with them were nonexistent as of August 2001. Consequently, in this project, we have tried to make what might be called "reasonable conjectures" about the associated costs, for comparison with the resources potentially available to meet these costs.

Second, it is difficult, as well as unnecessary for our present purposes, to distinguish between ESDP/RRF requirements and those that have been separately undertaken by the four countries for enhancing and modernizing their capabilities as part of the DCI within NATO. From the standpoint of claims on defense budgets, whether the capabilities in question represent expanding airlift or C^4I capabilities within the DCI/NATO framework or whether the enhanced capabilities are assigned to the ESDP, all of these costs represent requirements for defense resources from the contributing countries—principally, Germany, France, the United Kingdom, and Italy.

Third, the costs we are focusing on relate solely to military investment—that is, to military procurement and RDT&E. We have not considered O&M costs, which might entail an additional cost of 30 to 50 percent of the capital costs associated with equipping the enhanced ESDP force. We have neglected O&M for several reasons: European military establishments are in various stages of moving from conscript to volunteer forces. While this shift will heavily impact personnel (and hence O&M) costs in European defense budgets, the timing and magnitude of this change involve complications that extend well beyond the EMPEC study. The ESDP/RRF will consist not of new forces, but rather of previously existing manpower with new capabilities. The enhanced capabilities will result from new investments (i.e., procurements and RDT&E) to equip these forces and further training to enable them to utilize the systems effectively. Consequently, there may be greater fungibility and flexibility in O&M budgets, enabling them to be shifted from the "old" to the "new" ESDP forces, than there is in budgets for procurement and RDT&E. For these reasons, the approaches we have followed to assess ESDP costs are confined to investment costs.

APPROACHES FOR ASSESSING COSTS

To arrive at "reasonable conjectures" concerning the relevant incremental investment costs and prospective burdens on European defense budgets, we have applied four approaches:

- A "bottom-up" approach that focuses on the major system acquisitions that the force is expected to require and the prices and capital costs of these acquisitions.
- A generalized "top-down" approach using the procurement and RDT&E costs per member of the U.S. armed forces as a basis for estimating the capital costs of the 60,000-man ESDP/RRF.
- An approach that uses a rough estimate of the capital costs associated with a U.S. Marine brigade as a plausible building block for the ESDP force, upscaling it to the level of the 60,000-man ESDP/RRF force.

- An approach similar to the preceding one, but using instead a RAND costing model with a U.S. Army assault division as the modular building block.

"Bottom-Up" Approach

This approach focuses on the major system components presumed to be necessary to achieve the targeted military capabilities. As noted above, nothing exists to date that is specific enough to permit drawing up an approximate TO&E for the proposed high-tech, modernized, rapidly deployable, and interoperable force. Instead, there are verbal statements of "Headline Goals" for a force that would be "deployable within 60 days," that would have a manpower level of 50,000 to 60,000—approximately equivalent to an army corps along with its air and navy components—and that would be sustainable for at least a year of operation in the field.[1] It is worth noting that the issue of precisely what capabilities the RRF should be endowed with has not yet been resolved within the EU. Our cost estimates, while conjectural, implicitly assume the "high end" of the Petersberg tasks, which the French have strongly advocated, though as yet without full endorsement by the Union.

Verbal descriptions of the qualitative requisites associated with this force also include a European Air Transport Command to expand airlift, enhanced sealift capabilities, improved C^4I, and acquisitions of cruise missiles, GPF-guided bombs, UAVs, and missile defenses.[2] Without specification of the quantities of the systems to be acquired or upgraded, the cost estimation methods we have used remain conjectural.

Our cost estimates using the "bottom-up" approach draw on work by Colonel Franz Osinga of the Netherlands Air Force.[3] This work fo-

---

[1]Francois Heisbourg, "Europe for the Strategic and Ambition: The Limits of Ambiguity," *International Institute of Strategic Studies*, Summer 2000, pp. 5–15.

[2]James T. Thomas, "The Military Challenges of Transatlantic Coalitions," *International Institute of Strategic Studies*, 2000, pp. 68–70.

[3]Colonel Osinga's estimates, which he has characterized as "rough" and "somewhat arbitrary," purport to "consist of equipment that countries would not procure for national needs nor for NATO requirements." Therefore, in principle and conceptually, this set of requirements is *less* than what we have argued above would represent

cuses on major system acquisitions to provide strategic airlift, enlarged AAR capacity, early warning, electronic and signal intelligence, communication capabilities, theater surveillance, and CSAR helicopter forces. Table 4 summarizes these estimates.

The total capital costs estimated in Table 4 range from $37 billion to $47 billion. These estimates should probably be regarded as lower-bound figures, for two reasons: (1) they neglect RDT&E costs associated with adapting these procurements to European force and organizational circumstances, and (2) they explicitly omit the NATO/DCI acquisitions which, as we suggested earlier, represent additional claims on defense budget resources for the European NATO members as well as for the EU/NATO and non-NATO members.

Table 4

Preliminary Estimates of ESDP-Related Investment Costs (billions of U.S. 2000 dollars)

Exclusively ESDP–Related Investments	Number/Unit Cost	Total Investment Costs
Strategic air transport	225 (C–5 or equivalent)/0.1	22.5
Enlargement AAR capacity	20–30/0.15	3–4.5
SEAD capacity	20–40/0.1	2–4.0
EW capacity	8–12/0.2	1.6–2.4
ELINT/SIGINT	4–8/0.5	2–4.0
AGS capability	4–8/0.35	1.4–2.8
All-weather strategic surveillance	12/0.1 (plus 4 Predator UAVs)	1.2 plus 1.0
All-weather theater surveillance	10/0.15–0.25 per strategic UAV	1.5–2.5
CSAR capability	12–24/0.1 per CSAR helicopter	1.5–2.5
Total		36.7–47.4

SOURCES: Unofficial estimates by Colonel Franz Osinga, Netherlands Air Force, of additional procurement costs required to meet "the most demanding Petersberg task." See *De Militaire Staat van de Europeese Unie,* Clingendael Research Paper, Clingendael Institute of International Relations, The Hague, May 2001.

---

aggregate claims on additional procurement and RDT&E resources. The latter would entail NATO/DCI requirements as well. Colonel Osinga's rough estimates, which reflect his own thinking and do not represent any official or unofficial source, were made in connection with the European Workshop that we conducted as part of the EMPEC project at the end of February 2001 at the Netherlands Institute of International Affairs at Clingendael in The Hague.

"Top-Down" Approach

This approach draws on work by Huber and Schmidt,[4] and the resulting estimates are at least as conjectural as the preceding ones. The approach begins with an estimate of the annual expenditure per U.S. soldier for new military investment and RDT&E.[5] We have raised the Huber-Schmidt figure of $62,900 per year per soldier in 1998 to $65,400, to allow for modestly increased prices for military procurements between 1998 and 2000. These annual investment expenditures represent additions to the previously accumulated stock of military capital. If we make the arbitrary but probably conservative assumption that the requisite stock of relatively modern and advanced military capital contributing to the enhanced capabilities of U.S. forces has been built up over a prior period of, say, six or eight years (net of depreciation), then equipping the 60,000-man ESDP force with equivalent military capital would cost between $23.5 billion [= (65,400)(60,000)(6)] and $31.4 billion [= (65,400)(60,000)(8)].

These estimates implicitly assume procurement and RDT&E costs for the ESDP force that would constitute entirely new outlays, excluding any allowance for existing procurement and RDT&E levels in the four countries. Huber and Schmidt's estimate of the average current investment outlays per soldier in the four EU countries amounts to $30,650 in U.S. 2000 dollars. We have made no allowance for these ongoing military investments, on the basis of the arguable premise that much of these outlays would continue without contributing appreciably to the high-technology, advanced new systems that the rhetoric accompanying the ESDP/RRF implies.

Estimates Based on a U.S. Marine Expeditionary Brigade

The third approach to estimating the procurement and RDT&E costs of the ESDP/RRF is to analogize it to a hypothetically equivalent U.S. Marine expeditionary force. Specifically, we have used as a building

---

[4]Reiner Huber and Bernhard Schmidt, *The Challenge for Defense Reform in Europe: Conclusions from an Analysis of Defense Budgetary and Conceptual Constraints on European NATO Forces,* The Potomac Foundation, McLean, Virginia, 2000. Dr. Huber is Professor of Applied System Science at the Federal Armed Forces University in Munich, Germany, and Bernhard Schmidt is a captain in the German Army.

[5]Ibid., p. 17.

block a U.S. Marine expeditionary brigade (MEB) and the associated amphibious shipping and airlift to deploy it to a distant theater.

An MEB consists of 13,500 Marines, whereas a Marine expeditionary force (MEF) consists of three MEBs and a large command element. The main equipment costs of an MEB are shown in Table 5.

The total capital costs of the MEB shown in Table 5 come to $11.8 billion in U.S. 2000 dollars. Using the EU's target figure for a 60,000-man RRF implies upscaling the MEB figure by a factor of 4.44 [= (60,000/13.500)]. The result is an estimate of total capital costs of $52.4 billion [= (11.8)(4.44)] for the ESDP force.

Table 5

Equipment Costs for a Marine Expeditionary Brigade (MEB) (billions of U.S. 2000 dollars)

Equipment	Unit Cost	Total Procurement Cost
Naval ships		
2 LSD–41	0.7	1.4
2 LHD–1	1.8	3.6
2 LPD–17	0.9	1.8
3 MPS (RO/RO)	0.35	1.1
Total Navy ship costs		7.9
Marine equipment		
Ground forces (108 AAV, 25 LAV, 14 M1A1, 54 155 mm artillery, 136 trucks, etc.)	—	0.68
Marine air group		
37 F/A–18	28.1	1.0
16 AV–8B	23.7	0.38
5 EA–6B	52.0	0.26
24 MN–22	40.1	0.96
16 CH–53E	26.1	0.42
18 AH–1W	10.7	0.19
Miscellaneous (C130, UH–1M)		0.12
Total Marine equipment costs		3.99
Total MEB equipment costs		11.88

SOURCES: Unofficial estimates, courtesy of Commander Pete Dawson, USN, based on data from the Center for Defense Information, OPNAV; Marine Corps web site; records of foreign military sales, May 2001.

Estimates Based on a Mobile Advanced U.S. Army Division

The fourth approach is to estimate the capital costs of the ESDP force based on the equipment and associated acquisition costs of a U.S. Army air assault division, as derived from RAND's cost model for this unit. Table 6 shows these costs in 1997 U.S. dollars.

Converting the 1997 dollar figures to 2000 dollars gives a total capital cost estimate for the assault division of $3.48 billion. The equipment costs of the necessary support "tail" for the air assault division add

Table 6

Equipment and Materiel Costs of an Army Air Assault Division (billions of 1997 U.S. dollars)

Major Equipment Items	Total Procurement Cost
Aircraft	1.67
72 AH64 attack helicopter	
32 OH58 observation helicopter	
129 UH60 utility helicopter	
48 CH47 cargo transport helicopter	
Missiles	0.05
54 light artillery (105 mm)	
27 air defense artillery	
Weapons and truckload combat vehicles	0.07
569 5-ton, 551 2.5-ton, 2,118 1.25-ton and HMMWV	
Other procurement	
Tactical and nontactical vehicles	0.17
Telecoms and other communications	0.42
Other support equipment	0.44
Other equipment and medical supplies	0.35
Total integrated provisional equipment costs	3.27
Air transport for rapid deployment	21–42
210 C–5 transports (@ $100 million–$200 million)	

SOURCES: "Forces Cost Models," *U.S. Department of the Navy*, Falls Church, Virginia, 1997; air transport requirements courtesy of James Bigelow, RAND, Santa Monica, California, 2001.

approximately another 10 percent to the capital costs of the division itself, bringing the total capital costs for the division to $3.83 billion.[6] The manpower size of the assault division is 16,358; upscaling the assault division's costs to the intended 60,000-man ESDP force would, in turn, increase the capital costs of the ESDP force to $14.13 billion, to which must be added the capital costs of the air and sea transport required to enable the ESDP force to be rapidly deployable to remote areas. If we use the strategic air transport estimates from Osinga's calculations[7] of approximately $22.5 billion or those of our RAND colleague James Bigelow (which range from $21 billion to $42 billion), the result is a total capital cost estimate based on the reconfigured assault division of between $35 billion and $56 billion.

In sum, using the four approaches, we estimated capital costs for the ESDP force at between $24 billion and $56 billion.[8]

POTENTIAL SOURCES OF FUNDING

There are four possible sources of funding to meet the substantial procurement and RDT&E costs of the proposed ESDP force. Implicitly, there is a fifth source consisting of possible combinations among the four.

The first source is additional resources for military spending and military investment above and beyond the current flows of such resources. These additions could be generated by economic growth and justified by a reappraisal of a new range of security threats and

---

[6]The approximation for the additional equipment costs of the support tail covers a range from 6 percent to nearly 15 percent, depending on the distance to which the assault division is to be deployed.

[7]See p. 28 above.

[8]In a previous RAND study, Berman and Carter estimated the investment costs of a "robust" European projection force at between $9 billion and $25 billion in 1991 dollars (M. B. Berman and G. M. Carter, *The Independent European Force: Costs of Independence*, RAND, MR–178–AF/A/ASD, 1993). Reflating these estimates into 2000 dollars results in a range of $11 billion to $29 billion. The range of the Berman and Carter estimates depended largely on the projection distance in the several deployment scenarios they considered. The principal conceptual difference between their estimates and those presented in this study relates to investments in equipment and systems *additional* to the air and sea lift requirements we have associated with the RRF, e.g., C^4I, UAV, EW, etc. (see Table 4 above).

contingencies not only in the Balkans, but in areas of concern to the EU that are geographically more remote from Europe, including Africa, the Middle East, and Southwest Asia. This perspective reflects our focus on economic growth in the four principal EU/European NATO member countries and their ability to generate some additional resources for military spending and military investments as a consequence of economic growth. However, as noted earlier, most of the European *cognoscenti* with whom we have discussed this subject remain skeptical that this assumption is realistic in the context of European political and economic constraints. Instead, they argue that potential military claims on additional resources generated by economic growth rank well below competing claimants, including debt reduction and tax reduction, as well as social and "green" programs. In support of this view, they cite the fact that German defense spending has decreased (although the Germans aver that they will offset this by "investing more wisely" in the future) and that there have been evident shortfalls in meeting both DCI and ESDP commitments.

Apart from resources that might be generated from economic growth, a second source could, in principle, be reallocations in existing government budgets from nondefense to defense purposes—for example, from public subsidization of agriculture or from various entitlements. It goes without saying that our European interlocutors were even more emphatic in rejecting this option as unrealistic, so we omit it from the discussion that follows.

A third potential source of funding is the reallocation of existing military spending and military investments from their somewhat "backward-looking" focus—for example, on heavy tanks, artillery, surface ships, etc., all relating to World War II imagery—to a more forward-looking, high-technology, C^4I, air-mobile, and deployable set of capabilities. Although several European experts were receptive to this possibility, others (as well as the authors of this report) are doubtful of the extent to which such reallocations from the current pool of defense resources constitute a realistic prospect.[9] Resistance

---

[9]A French Army colonel, in his technical review of this study, corroborated the authors' doubts with this observation: "About the reallocation, I don't see what can be done to allow [for a] shift of money of this magnitude. *The European armies are very rigid organizations*" [emphasis added by the authors].

to such reallocations springs from two sources: First, the established types and channels of procurement for the military services typically represent important interests and organizations within both the militaries and the defense industries of the four countries; second, notwithstanding the growth of integrating institutions and attitudes within the EU, none of the principal countries is likely to forgo its historical commitment to territorial protection, as well as to protecting its individual and separate national interests outside of territorial boundaries. Established patterns and types of military procurement reflect these historic commitments, and hence these countries are likely to be resistant to major changes.

The fourth source of potential funding for the ESDP could materialize through various measures designed to liberalize and consolidate the existing inefficient segmentation of European military procurement and defense industries. Proposals for moving in this direction have been advanced for many years by several European scholars and analysts, in particular, Professor Keith Hartley, Lord John Roper, and Timothy Garden in the United Kingdom.

Among the four member states, support for independent national defense industries has resulted in duplication of costly R&D programs, relatively short production runs, and small national orders. For example, Hartley has estimated that if procurement were on a competitive and supranational basis, with open, liberalized, and competitive bidding, the resulting annual savings in defense procurement would range from 10 percent to 17 percent, or from $10 billion to $15 billion annually.[10] In effect, if such measures were successful, the result would be "more bang per Euro," a de facto equivalent to providing additional resources for meeting the increased investment costs of the ESDP. While some progress in this direction has been made and probably will continue, hardly less skepticism is warranted concerning this potential source of additional funding as pertains to the preceding sources.

Finally, various combinations of these sources of funding may be possible and indeed may be more realistic than exclusive reliance on any one individual approach.

---

[10]Keith Hartley, "A Single European Market for Defense Procurement," University of York, unpublished briefing, October 2000.

Chapter Five

POLITICAL AND STRATEGIC ISSUES: CONCLUDING OBSERVATIONS

SOURCES OF FUNDING AND ESDP COSTS

Based on the assumed capital costs—including procurement and RDT&E—for the fully operational ESDP force of between $24 and $56 billion, Table 7 summarizes the potential sources of funding to meet these costs.

As the first column of Table 7 indicates, if the existing allocation of resources for military investments is unchanged, additional resources generated by economic growth will be far below the estimated capital costs of the ESDP force over the next ten years. However, if at least one-third of the present allocations for annual military investments is shifted and reallocated to the procurement

Table 7

Sources of Funding (billions of U.S. 2000 dollars)

	Incremental Resources for Military Investment	Reallocations (by One-Third) from Annual Military Investments	Savings from Consolidation of the Defense Industry and the European Defense Market
2001–2003	5	20–30	6
2004–2007	18	30–40	10
2008–2010	22	20–30	

SOURCE: See Table 3 and text, pp. 32–34.

needs associated with the ESDP/RRF, then by 2010 sufficient funding would be available to meet the estimated capital costs of the force.[1] Finally, as the third column of Table 7 indicates, savings from increased efficiency in the defense industry and from movement toward a single defense market in the EU would contribute additionally, if modestly, to defraying the necessary capital costs.

Four principal conclusions emerge from this analysis:

1. Meeting the capital costs of the ESDP/RRF by the planned 2003 target year is unlikely.
2. Assuming that incremental investment resources are available, but without substantial reallocations from existing military spending and military investments, the requisite capital costs for the ESDP/RRF cannot be met until the end of the decade.
3. However, with such reallocations, the necessary capital costs can be met by 2007.
4. If combinations among the several funding sources can be accomplished, realization of the goal can be further accelerated.

To move more aggressively in this direction will require overcoming organizational inertia and established service and industry interests. The political will to accomplish this may well be forthcoming, although thus far the rhetoric behind the ESDP has proceeded far more rapidly than has the reallocation of needed resources.

In sum, we conclude that the United States and NATO have less reason to worry that the EU will acquire the enhanced military capabilities required for the ESDP/RRF than that it will continue to produce descriptive rhetoric without the resources necessary to acquire and support them.

---

[1]One of our interlocutors (the French colonel referred to in footnote 9 of Chapter Four) expressed the opinion that our estimate of $20 billion to $30 billion is too high. Instead, he opined, "A reallocation of $10 billion per period will be great and for me very unlikely."

POLITICAL AND OTHER NONECONOMIC ASSETS OF THE ESDP

As noted earlier, our EMPEC project has focused on defense economics, although we recognize that political, strategic, and other noneconomic issues represent a large and probably dominant share of the ESDP terrain.[2] For example, a central concern from the U.S. point of view is whether the ESDP, as the foreign and security "pillar" of the EU, will help, hinder, or not affect the cohesion and effectiveness of NATO in the years to come.

The answer to this question is clouded by the sharply differing views of the United Kingdom and France, with Germany and Italy somewhere in-between and the other members of the EU spread across the spectrum. For example, as noted in Chapter One, Britain's Defense Secretary recently asserted Britain's intention "to ensure that any proposals for a rapid reaction capability inside the EU are completely consistent with the NATO planning process."[3]

On the other hand, the head of France's Armed Forces, General Jean-Pierre Kelche, subsequently stated, "If the EU works properly, it will start working on a crisis at a very early stage, well before the situation escalates. NATO has nothing to do with this. At a certain stage, the Europeans would decide to conduct the military operations. Either the Americans will come, or not."[4]

The complexity of these political and other issues is partially conveyed by the substantial, but incomplete, overlap among the respective memberships of the EU, the EMU, and NATO—the so called EU 15, EMU 12, and NATO 19 (see Figure 6).

As Figure 6 indicates, three of the four countries we have focused on (Germany, France, and Italy) are members of all three organizations. The United Kingdom is a member of two, but not of the EMU; the

---

[2]For an illuminating treatment of this broader context, see Robert E. Hunter, *European Security and Defense Policy: NATO's Companion—or Competitor? An American Perspective*, RAND (forthcoming).

[3]See p. 2 above.

[4]*The Weekly Telegraph*, electronic file, March 28, 2001.

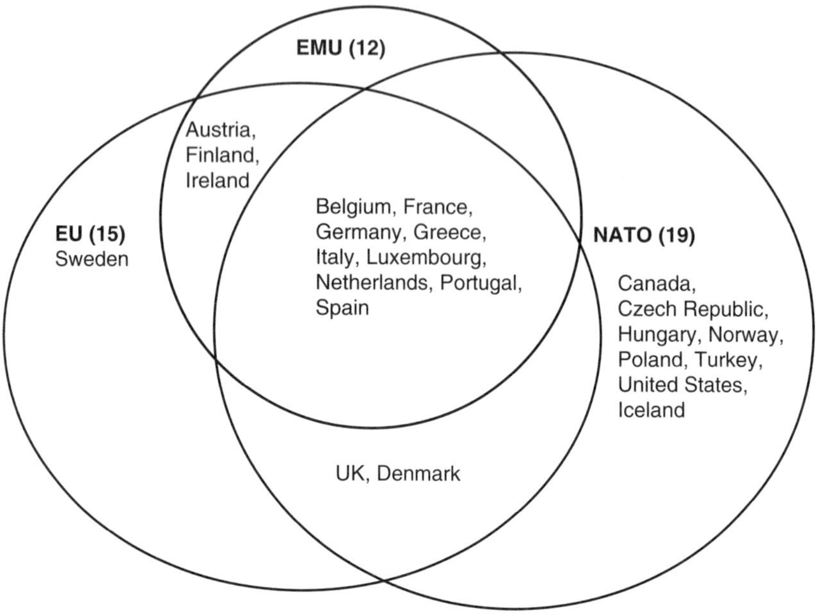

Figure 6—NATO 19, EU 15, and EMU 12

Czech Republic, Hungary, Norway, Poland, Iceland, Turkey, Canada, and the United States are members of NATO, but not of the other two entities; Sweden is a member of the EU, but not of the others; and Ireland, Finland, and Austria are members of the EU and the EMU, but not of NATO. Reaching consensus within each organization, let alone between them, is often a formidable challenge.

In addition to the complications presented by these partly overlapping and partly divergent memberships, the effect of the ESDP on NATO—and on U.S. interests more broadly—will depend to a considerable extent on how the ESDP is implemented. For example, the ESDP's congruence and complementarity with NATO will be affected by whether it maintains a planning structure separate from or integrated with that of NATO, i.e., whether its command structure is linked with that of NATO so that CINC/ESDP is also located within the NATO structure, perhaps as Deputy SACEUR. Actually, this command relationship appears to be moving in a different direction, with the preliminarily designated acting commander of the still nonexistent ESDP force a general from Finland, rather than someone from one of the established NATO countries within the EU.

How the organization of the ESDP develops, how the force is trained, and how its strategic doctrine evolves will significantly affect the extent to which the force contributes to and complements U.S. strategic interests. If, for example, the doctrine and training of the EU force place primary emphasis on autonomous operations and disconnection from NATO, the effect may be to weaken NATO and even bring into question its reason for being and hence its longevity. If, on the other hand, the ESDP's equipment, training, and doctrine evolve in ways that emphasize linkages, collaboration, interoperability, and joint operations with the United States and NATO, there may be significant synergies between development of the ESDP force and U.S. strategic interests.

This outcome will apply especially to "out-of-area" contingencies in Africa, the Middle East, and Southwest or East Asia. In these areas, ESDP forces might plausibly operate jointly with U.S. forces, avoiding the constraint that Article 5 and Article 6 of the NATO Treaty impose on the military operations of NATO forces outside the treaty area.[5] Indeed, other RAND work envisages precisely this possible mode of future collaboration and burden-sharing between EU and U.S. forces.[6]

---

[5]Article 4 of the Treaty does provide a way of avoiding this constraint. Article 4 states: "The Parties will consult together whenever, in the opinion of any of them, the territorial integrity, political independence or security of any of the Parties is threatened." Robert Hunter, formerly U.S. Ambassador to NATO, in his technical review of this study, avers that "Article 6 in no way limits the *discretionary* role of NATO forces" acting beyond the treaty's geographic area. In his view, if the EU countries as well as the United States were "unanimous in wanting to act beyond Europe," it would be "virtually inconceivable" that the resulting joint operations would not become a NATO operation. For a somewhat different view, see James Thomson, "A New Partnership, New NATO Military Structures," in David Gompert and Stephen Larrabee (eds.), *America and Europe,* Cambridge University Press, 1997, pp. 83–85.

[6]Stephen Larrabee, Edward Warner, and Stuart Johnson, *Alternate U.S. and EU Defense Strategies: Political and Military Implications,* RAND, AB-501-OSD, 2001. The authors acknowledge, however, that this model would be a "hard sell" and that Europe "would probably resist" it.

Appendix A

MODEL AND METHOD

THE FORECASTING MODEL

The forecasts of economic and military trends presented in this report are based on a hierarchically linked model in which (1) GDP (gross domestic product or gross national product[1]) is estimated from a simple Cobb-Douglas-Solow production function whose arguments are the capital stock, employed labor, and the productivity of capital and labor; (2) military spending is estimated as a specified, sometimes varying, proportion of GDP; and (3) military investment is estimated as a specified, sometimes varying, proportion of military spending.

Choice of this model is based on its commendable transparency, its convenience for calculation purposes, and its modest and tractable data requirements compared with, say, input-output models, translog production functions, or time-series regressions. The method used to derive military spending and military capital estimates was selected for similar reasons of tractability, simplicity, and transparency.

The model used in the forecasts consists of five variables—GDP, employed labor, nonmilitary capital, military spending, and military investment, each of which carries a time subscript—and six parameters—annual employment growth, annual growth of nonmilitary

---

[1]The estimates we present are for GDP. The accounting relation between GDP and GNP is GDP = GNP – net factor income from abroad.

capital, total factor productivity (τ) representing the annual rate of technological change, the labor share in GDP (α), the proportion of GDP devoted to military spending (γ), and the proportion of military spending devoted to military investment (π). The model is first applied to data covering the period from 1985 through 1999; the parameters are estimated by calculating their mean values and variances over this period and, in some cases, by regressing the parameter values in each year on time to determine trends.

We then use these parameter values and appropriate values for the input variables based on the authors' explicit judgments about whether these trend values are likely to persist or why they may be expected to change and by how much. These judgments are described and reflected in the discussion in Chapters Two and Three.

The model, summarized below, was used for each country, together with adjustments and elaboration to allow for data problems or other country-specific circumstances.

$$Q = (e^{\tau t}) \bullet L^{\alpha} \bullet K^{(1-\alpha)} \quad (1)$$

$$MS_t = \gamma GDP \quad (2)$$

$$MI_t = \pi MS_t \quad (3)$$

where

Q = GDP

τ = rate of technological change (total factor productivity)

t = years covered in the projections, beginning with 1994

α = labor share in GDP

L = labor input in each year

K = capital input in each year

MS_t = military spending in year t

γ = proportion of GDP devoted to military spending

MI_t = military investment in year t

π = proportion of military spending devoted to procurement of equipment and RDT&E

In Equation (1), the civil capital inputs (K) and labor inputs (L), along with their corresponding growth rates, were estimated for each country. The capital input, K, for each year was calculated by adding each year's net new investment to the previous year's civil capital stock.[2] Note that this depreciation rate on the civil capital stock is not necessarily the same as the depreciation rate on the military capital stock.

Equation (1) can be expressed in a form that is useful for our forecasts by taking the logarithmic derivatives of the variables with respect to time. The result is

$$\dot{Q}/Q = \tau + \alpha(\dot{L}/L) + (1-\alpha)(\dot{K}/K) \tag{1a}$$

Equation (1a) stipulates that the rate of growth in GDP is equal to the annual growth of total factor productivity (technological progress) (τ), plus the rate of growth in employment multiplied by the share of labor income in GDP (α), plus the rate of growth in the capital stock multiplied by the share of capital income in GDP ($1 - \alpha$). The rate of growth in total factor productivity in each country in recent years can be estimated from the known values of the other variables in Equation (1a).

The labor and capital income shares, (α) and ($1 - \alpha$), respectively, are based on the respective data and experience of each country, as indicated in the source references for Table 1.

Similarly, estimates of the parameter γ, representing the share of GDP devoted to military spending, are calculated from each country's average share in recent years, combined with explicit judgments by the authors, as noted above.

Annual military investment comprises military procurement plus expenditures on RDT&E. Our estimates of the parameter π, representing the share of military spending devoted to military investment, are based on each country's average share in recent years as observed in the 1985–1999 trend period.

---

[2]Where no base-year capital stock figure was available, we assumed an aggregate capital-output coefficient of 4 and an annual depreciation rate of 8 percent.

METHODOLOGICAL COMMENTS

Conversions to U.S. Dollars

Most of the data we used (see Tables 1 through 3 above, pp. 10, 11, 12, and 19) in analyzing historical trends over the 1985–1999 period were expressed in the primary and secondary sources in terms of "own currencies," that is, deutsche marks, francs, pounds, or lira, in current or constant prices. Typically, the constant-price series were expressed in 1995 prices in the original sources. Conversions to U.S. 2000 dollar equivalents posed a choice between two methods: (1) reflating the 1995 own-currencies series to 2000 prices, using the respective GDP deflators of each of the four countries and then converting to U.S. dollars through the nominal exchange rates between the U.S. dollar and each of the four currencies prevailing at the end of year 2000; or (2) converting the own-currencies 1995 series to U.S. dollars at the nominal exchange rates prevailing in 1995, and then reflating the 1995 dollar series to 2000 U.S. dollars, using the U.S. GDP deflator.

The two methods would yield convergent estimates *if* the parities between the dollar and the four European exchange rates (three of which have been tied to the Euro since January 1, 1999) moved in precise accordance with movements in their respective GDP deflators. In fact, this theoretically plausible pattern did not occur. Inflation rates in the U.S. and the four European countries were similarly low, but the four European currencies weakened (10 percent for the British pound and 15 to 20 percent for the Euro-tied currencies) against the dollar between 1995 and 2000. The explanation for this divergence, of course, lies in the direction and magnitude of capital flows into U.S. dollar assets.

The calculations presented in this report were made using the first of the two methods, on the grounds that this method yields more realistic estimates of current values of the relevant variables than would the second method.

Capital-Stock Estimation

None of the original data sources provided estimates of the stock of productive capital in each of the four countries. Consequently, we used two different methods to arrive at base-year capital-stock estimates. These, in turn, enter into calculations of the rate of growth of the capital stock ($\dot{K}/K$) in the production-function model described above.

The method applied to Germany and Italy assumed their respective capital stocks in 1969 to be six times capital formation in that year, with new capital formation added in each subsequent year and annual depreciation of 8 percent applied to existing capital stock subtracted in each subsequent year. The method applied to the United Kingdom and France assumed the capital stock in 1985 and subsequent years to be equal to four times GDP *plus* annual capital formation and *minus* 8 percent depreciation of the existing capital stock. With reasonable assumptions about annual rates of capital formation, the first method yields a capital stock estimate for 1985 and subsequent years that is somewhat more than three and one-third times GDP in 1985—close to the assumed average capital-output ratio of four, which was assumed in the second method.

Thus, for purposes of deriving rough estimates of the respective capital stocks in the base years, the two methods are nearly equivalent.

SELECTED DEMOGRAPHIC DATA FOR GERMANY, FRANCE, THE UNITED KINGDOM, AND ITALY

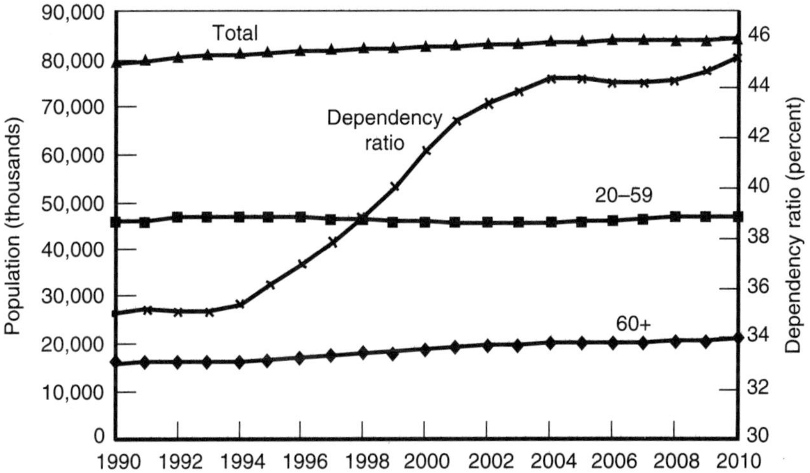

SOURCE: U.S. Bureau of the Census, *International Database,* 2000. Figures for 1990 interpolated from census data.

Figure B.1—Germany: Demographic Data

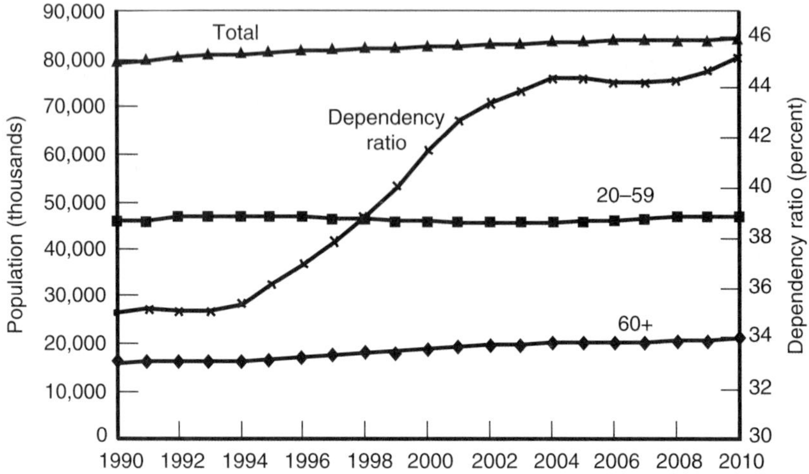

SOURCE: U.S. Bureau of the Census, *International Database*, 2000.

Figure B.2—France: Demographic Data

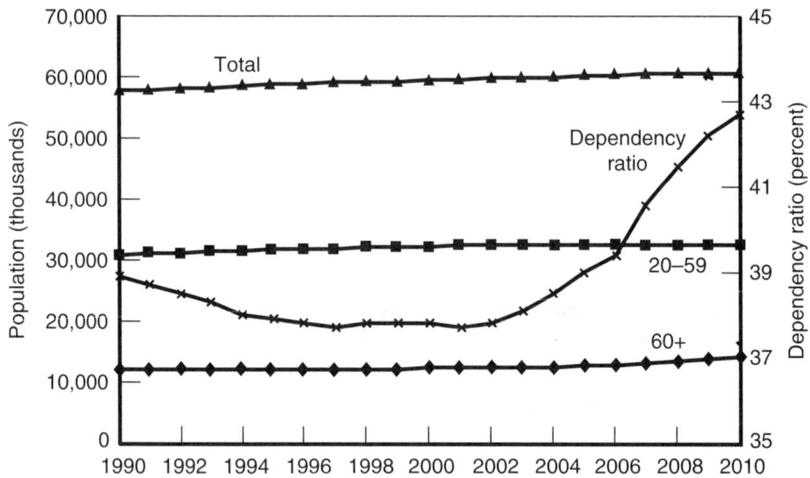

SOURCE: U.S. Bureau of the Census, *International Database*, 2000.
Figures for 1990 interpolated from census data.

Figure B.3—United Kingdom: Demographic Data

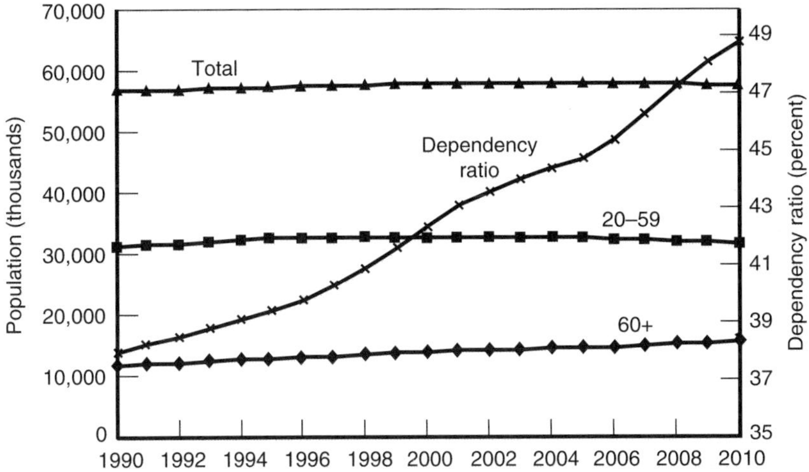

SOURCE: U.S. Bureau of the Census, *International Database*, 2000. Figures for 1990 and 1991 interpolated from census data.

Figure B.4—Italy: Demographic Data